S0-CLV-741

DATE DUE			

CLOCKS IN THE ROCKS

Learning About Earth's Past

Patricia L. Barnes-Svarney

—an Earth Processes book—

ENSLOW PUBLISHERS, INC.
Bloy St. & Ramsey Ave.   P.O. Box 38
Box 777   Aldershot
Hillside, N.J. 07205   Hants GU12 6BP
U.S.A.   U.K.

Copyright © 1990 by Patricia L. Barnes-Svarney.

All rights reserved.

No part of this book may be reproduced by any means without the written permission of the publisher.

Library of Congress Cataloging-in-Publication Data

Barnes-Svarney, Patricia L.
  Clocks in the rocks: learning about earth's past / by Patricia L. Barnes-Svarney.
    p. cm.—(An Earth processes book)
  Includes index.
  Summary: Discusses the divisions of time used by geologists to record Earth's history and includes information on fossils, radioactive dating, and prehistoric life.
    1. Geological time—Juvenile literature. [1. Geology.]
  I. Title. II. Series.
  QE508.B355 1990
      551.7—dc20   89-7698
                                      CIP
                                      AC

ISBN 0-89490-275-X

Printed in the United States of America
10 9 8 7 6 5 4 3 2 1

Illustration Credits:
Frank Carpenter, p. 6; Marie Morisawa, pp. 37, 39, 49; Museum of Natural History, University of Kansas, Lawrence, p. 42; National Aeronautics and Space Administration, pp. 21, 24; National Park Service, photo by Richard Frear, p. 8; **Natural History** Museum of Los Angeles County, pp. 53, 58; Peabody Museum, Yale University, pp. 15, 46; David W. Tuttle, pp. 10, 35; United States Geological Survey, pp. 7, 26, 30.

Cover Photo: United States Department of the Interior, National Park Service.

Contents

1 Fossils Galore 5
2 Clocks in the Rocks 14
3 Precambrian: The Beginnings 23
4 Paleozoic: Signs of Life 32
5 Mesozoic: Middle Ages of Time . . 41
6 Cenozoic: The Earth Today 50
   Glossary 59
   Further Reading 62
   Index 63

1
Fossils Galore

The ant worked hard to carry his burden to the nest. Then he quickly scampered up a pine tree in search of more food. A small drop of resin seeped from the pine tree, and the ant became caught in the sticky substance. Trapped in the pine sap, the ant soon died. This event happened more than 30 million years ago, and today we see the ant and other ancient insects in amber, a rock made of hardened pine sap resin. All the creatures are preserved as though they had been working yesterday!

The ant is just one of the many creatures that have roamed the earth over time. There were many other animals and plants. Strangely shaped plants called *Cooksonia* lived close to 400 million years ago. Animals like the *Brachiosaurus*, a dinosaur that lived more than 80 million years ago, was one of the largest land creatures of all time. All together, millions of species of plants and animals have lived at one time or another on the surface of the earth.

What are fossils? Any remains or traces of an ancient plant or animal are called fossils. Fossils can be the remains of the durable part of a plant or animal, such as wood, teeth, shells, or bones. Or fossils can be only an imprint, such as the foot print of a dinosaur. Most fossils have been found in solid rock, but they have also been discovered in sticky tar pits, loose sand, and frozen in ice!

Scientists who study fossils are called paleontologists. Some of them specialize in certain types of fossils, such as the remains of dinosaurs or fern-like plants.

One of the earliest paleontologists was Herodontus, a Greek historian and traveler. After finding fossil seashells in Egypt, Herodontus correctly concluded that the sea had at one time covered lower Egypt. In the fifteenth century, Leonardo da Vinci, the great Italian painter and scientist, found fossils in the highlands of northern Italy. Despite opposition from other scientists of his day, Leonardo correctly concluded that the highlands were once covered by an ancient sea.

In 1802, Jean Baptiste Pierre Antoine de Monet, the French geologist, was one of the first to use the word "fossil" to describe the remains of an organism. Thomas Jefferson, the third president of the United States, was very intrigued by fossils. He sent William Clark (of the Lewis and Clark Expedition) to collect fossils in Kentucky—adding all sorts of specimens to the president's collection!

Fossils are usually found in sedimentary rocks—a type of rock that makes up about three-quarters of the land surface of the earth. As

The fly is in amber, a rock made of hardened resin from pine trees. The rock is more than 30 million years old.

the name suggests, these rocks are made up of sediments or particles of dirt and dust. Most commonly, the sediments are from rocks that were ground down, or eroded, by wind and water.

There are many types of sedimentary rock found around the world. Limestone is formed when sea organisms die and fall to the ocean floor. Sandstone is commonly formed from the erosion of sand from beaches. Mudstones and siltstones are formed by deposits of mud and silt from oceans, lakes, and rivers. Layer upon layer of sediment is deposited. After thousands of millions of years, changes in temperature and pressure turn the sediment to stone.

How does a fossil form? Take, for instance, a snail in a small spiral-shaped shell. One day, the snail dies and falls softly to the ocean floor. The ocean floor is very shallow and calm in this area. As the animal decays, sand, mud, and other sediments bury the shell.

Layer upon layer of sediment buries the creature deeper, cutting off the oxygen that causes decay. As more layers build up, the underlying layers harden into rock. After thousands of years, the snail will

A paleontologist carefully chips away rock surrounding a dinosaur bone.

become a fossil—the remains represented by the snail's shell in a hard rock.

To become a fossil, a creature should have several hard parts. The bones and antlers of an elk, the horny armor of a crab, or the circular shell of a chambered nautilus are only some of the types of hard parts that can eventually become fossils.

Not all fossils form in the same way. In Arizona, around 225 million years ago, the land was filled with lush vegetation—with trees as tall as 250 feet (76 meters)! In the Southwest, volcanoes were erupting, adding tons of the mineral silica to the air and water. As the trees died and settled into a nearby swamp, the wood was replaced by silica that had accumulated in the water. The Indians of the area appropriately called the rock "wood that turned to stone." Today the rock is called petrified wood—fossil wood found in the Petrified Forest National Park in Arizona.

Fossils also form when water and minerals seep through the tiny

Logs from the Petrified Forest National Park were once tall trees around 225 million years ago.

cracks of sedimentary rocks. Water often dissolves the bones or shells buried in a rock. When the bone or shell is gone, a cavity, or mold, remains. If the cavity is filled with minerals, a cast is formed.

A fossil can form by mummification. An amazing example was unearthed in northeastern Siberia in 1976. Prospectors searching for gold along the Kolyma River noticed a large object frozen in the muddied ice. Looking closer, the prospectors found the frozen remains of a baby mammoth—an ancestor of the modern elephant. The icy ground had prevented the mammoth from decaying. Scientists estimated that it was around 17,000 years old!

Other fossils are mere traces of animals left in mud or sand. These are called trace fossils. They can be compared to looking at a worm trail on soft sediment after a rainstorm: the worm is gone, but its movements can be traced by the trail left in the soft mud. In ancient sediments, most trace fossils were formed by feeding animals. Their trails are seen as a series of long, wavy lines in a rock.

There are some organisms that never became fossils. Soft-tissue animals and plants, such as the wings of a butterfly or thin seaweed, are too fragile to become fossils. Other organisms are crushed by the natural movements of the earth (such as in an earthquake) or destroyed by heat from a volcanic eruption.

Where do paleontologists and amateur fossil collectors look for fossils? Fossils are often found in limestone or shale quarries, where the sedimentary rocks are mined for buildings, roads, and walkways. Fossils are also found in coal mines and gravel pits, along road cuts and riverbeds—and even in someone's backyard!

Scientists have determined that many fossils are millions of years old. Such old ages are difficult to understand, especially since humans measure time in seconds, minutes, hours, days, and years. The earth revolves on its axis approximately every 24 hours and rotates around the sun in about 365 days. The planet Saturn rotates once around the sun in about 29 1/2 years. A human life span may cover 100 years. Some sea turtles can live up to 200 years. The giant redwood trees in

Brachiopod casts are formed when the fossil is replaced by a mineral—in this case, the shiny mineral pyrite.

Brachiopod molds occur when only the imprint of the shell remains.

California are often more than 3,500 years old; and some bristlecone pines live more than 4,500 years. But there are few creatures or plants that live to be more than 5,000 years old.

In order to keep track of fossil ages and the earth's long history, scientists have developed a special scale to keep track of time—the geologic time scale. It is divided into several sections including eras, periods, and epochs.

Each division on the time scale has a certain name. The names are based on a Greek or Latin word or named after the area in which a rock of that age was first found. The Cenozoic era was named after the Greek *kainos* ("recent") and *zoe* ("life"). The Cambrian period is from the Latin name for Wales. The Devonian period is named after rocks found near Devonshire, England.

Each era, period, and epoch represents a certain amount of time. And each fossil and rock layer falls somewhere along the geologic time scale.

But how do scientists know where a fossil or rock layer fits on the time scale? They do it by determining the relative or absolute ages of the fossil or rock. Relative age is determined by the position of a rock or fossil compared to other rock layers or fossils. Absolute age is found by measuring the radioactive decay of rocks or some fossils. Radioactive materials within rocks and fossils decay at varying rates. By using special measuring instruments, scientists can approximate the absolute age of the rock or fossil.

Relative age places earth events, plants, and animals in an orderly time sequence. The Grand Canyon of the Colorado River is a good example of relative time. Layer upon layer of rock represents a sequence of time—from the oldest rock layers at the bottom of the canyon to the youngest at the top. The relative age of a fossil found in the Grand Canyon is based on the position of other fossils around the original.

In the late 1700s, scientists had little knowledge about the relative ages of fossils. But eventually, two men helped to prove the idea that the earth's history could be interpreted by the rock layers and fossils: James Hutton and William Smith.

James Hutton, a Scottish physician and geologist, made two important contributions to the interpretation of geologic time. As he walked on the Isle of Arran in his native country, he noticed how the rocks seemed to lie in layers, like the leaves of a book. Hutton believed that these layers formed one on top of the other—making the bottom layers the oldest rock and the top layers the youngest rock.

Hutton also developed the doctrine of uniformitarianism, an important idea in interpreting geologic time: the present is the key to the past. He reasoned that the earth's geologic processes, such as rivers carving great valleys and volcanoes spreading ash and dust, have happened in the past and will happen in the future.

William Smith was an English surveyor and engineer. He devoted 24 years to mapping rock layers, or strata, around England. He spent so much time studying rock layers that his friends gave him the nickname "Strata"!

For years, Smith collected fossils in his native England. In the early 1800s, Smith noted that certain types of strata seemed to contain specific fossils. He had an idea. If he could find a certain type of rock layer, he should be able to find a certain fossil.

Smith was right. He soon developed the principle of biologic succession—the idea that specific life forms are unique at certain times in the earth's history. Smith's discoveries were important. Now scientists would use fossils to determine the relative ages of certain rock layers.

At the same time in France, Baron Georges Léopold Cuvier and Alexandre Brongniart, confirmed Smith's findings. They showed that certain types of fossils are found within definite rock layers. With this information, scientists developed a geological time scale to describe the many fossil-filled layers found on the earth.

Paleontologists have discovered a great deal about the earth's long past based on the study of fossils. The more than three-billion-year-old blue-green algae fossils are thought to be evidence of some of the first living creatures on the earth. The fossils of the mighty dinosaurs—reptiles that dominated many parts of the world for more than 100 million years—reveal that the giants suddenly died out around 65 million years ago. And 200-million-year-old cycad fossils are ancestors of today's plants of the same name.

Fossils that are fairly abundant and occur within a specific time interval are called index fossils. The fossil *Phacops,* a type of trilobite, is a fine example. *Phacops* are found in rock layers that are approximately 400 million years old. The animal lived during the Paleozoic era, between the Ordovician and Devonian periods. This type of trilobite did not live at any other time during the earth's long history—making it a perfect index fossil!

Sometimes taking a walk through geologic time is like visiting the present. The 10,000-year-old remains of a woolly mammoth look like the modern elephant. Or the strange fossils of the 400-million-year-old comb jellies look remarkably like today's jelly fish. Yet there are other fossils like the trilobites that are unique and are not found in modern times.

The ancestors of several ancient fossils exist today. Most of the "living" fossils are smaller organisms. In particular, foraminifera, rounded single-celled organisms of the oceans, have existed from the Cambrian period to today. Ostracods, small shrimp-like creatures, have been in existence for 450 million years.

As more fossils are found, scientists will change many of their ideas about the plants and animals of the ancient earth. Only a few years ago, scientists believed that the giant woolly mammoth had long hair. But recent studies of the mammoth fossils show that the animal had very short hair!

Looking at fossils and using the relative time scale has advanced the study of paleontology and the knowledge of the earth's past. But another way of measuring fossils and rocks has become important to the interpretation of geologic time: absolute time.

2
Clocks in the Rocks

It was an unexpected find. In 1896, Antoine Henri Becquerel, a French physicist, placed a seemingly innocent rock on a piece of photographic film. When he picked up the rock, the film was exposed in several spots—as if the rock had taken a picture of itself! Even when Becquerel wrapped the rock in opaque material, the photographic plate was exposed.

Becquerel called his discovery radioactivity. He had been exposing his photographic plates to uranium—a radioactive material. In 1898, Becquerel's discovery of radioactivity led Marie and Pierre Curie, two French physicists, to find two more radioactive elements: polonium, named after Marie Curie's native country, Poland, and radium.

Since then, more than 40 radioactive elements have been discovered. The rate at which many of these unstable elements decay, or become less radioactive over time, has helped scientists to better understand the absolute age of rocks and some fossils. Thus, more information about the age of rocks and fossils has been added to the geologic time scale.

Absolute age can best be described as the "exact" age of a rock or fossil in years. (Remember, relative age is the age of the rock in relation to other rocks around it.) Absolute age can be determined by measuring the radioactive decay of certain elements within a rock or fossil.

Rocks are composed of minerals, and various minerals contain certain radioactive elements. Minerals such as uraninite, zircon, and monazite often contain radioactive uranium and thorium. Micas and feldspars frequently contain radioactive potassium.

In addition to radioactive minerals, some fossils contain radioactive carbon. It is often found in bones, teeth, and wood.

The measurement of radioactive decay within rocks and some fossils has helped to determine their absolute age. But in order to understand absolute age, it is necessary to know something about radioactivity.

Everything on the earth is made up of molecules—tiny particles that cannot be seen with the naked eye. Each molecule has a certain number of atoms or elements. Water is one of the most common

 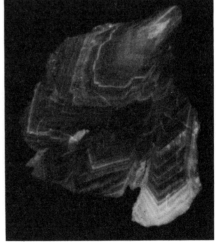

The rock samarshite is seen in regular light (left). When the same rock is laid on X-ray film, the radioactivity from the rock forms an image (right).

molecules, made up of two atoms of hydrogen and one atom of oxygen, or H_2O.

An atom can be further broken down into protons, neutrons, and electrons. Protons and neutrons make up the central core, or nucleus, of an atom, while the electrons spin around the nucleus. Each atom or element, whether it is oxygen (O), carbon (C), hydrogen (H), or nitrogen (N), has a certain number of neutrons and protons in its nucleus.

Radioactive atoms have a certain number of protons and neutrons in their nuclei. In the radioactive form of the atom uranium, there are 146 neutrons and 92 protons. The total of the neutrons and protons is 238. Scientists label this type of radioactive uranium U-238.

But the number of neutrons in other types of uranium nuclei, or other radioactive elements, can vary. Thus, the number of protons and neutrons for another radioactive uranium atom can total 235. This atom is called U-235. These different varieties of the same element are called isotopes.

During radioactive decay, an isotope naturally breaks down, releasing radioactivity. Certain measurable particles are thrown off, including alpha particles, gamma rays, and beta particles. Each particle or ray produces a certain amount of heat.

As the isotope breaks down, it turns into another, less radioactive isotope. Then that isotope breaks down into another isotope, and so on, until there is almost no radioactivity left in the rock.

Depending on the radioactive element, it may take a few microseconds or billions of years for an isotope to break down! The time it takes for half of the original isotope atoms to turn into the next isotope is called the half-life of the element.

Each isotope has a specific half-life. The element uranium is known because of its association with nuclear reactors and the atomic bomb. Uranium has two isotopes: uranium-238, with a half-life of 4.5 billion years, and uranium-235, with a half-life of 704 million years.

Rubidium-84 has a half-life of 51 billion years. Potassium-40's half-life is 1.31 billion years. Radioactive carbon, called carbon-14, has a half-life of 5,730 years. Aluminum-26 has a half-life of only 7 seconds. And beryllium-8 has a half-life of 0.0000000000000002 seconds!

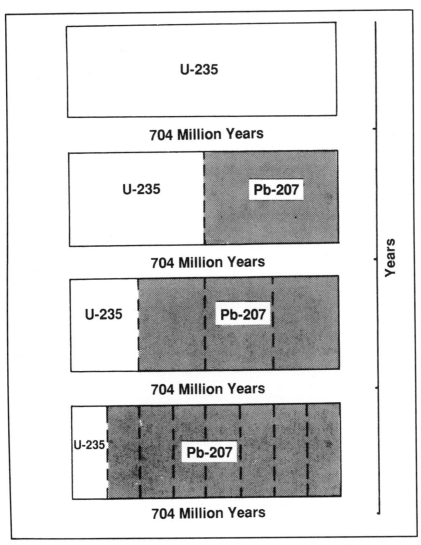

The radioactive decay of uranium-235 into lead-207 (Pb-207).

How does radioactivity develop in rocks? Magma, or liquid rock from deep below the earth's surface, often contains radioactive materials. As the magma rises to the surface and cools, minerals form. Some of these minerals contain radioactive elements.

As time passes, radioactive isotopes such as uranium decay within the mineral. By measuring the amount of radioactive decay, scientists are able to determine the absolute age of the rock.

One of the first scientists to use radioactivity to determine the age of rock was B. B. Boltwood. In 1905, Boltwood realized that the metal lead may be the end product of the radioactive decay of uranium. In 1907, by using special dating techniques, he presented new evidence that many types of rock were millions to billions of years old. Boltwood's pioneering work expanded the field of dating ancient rocks. Less than a century later, scientists still use radioactivity to determine the absolute ages of rocks.

Why do scientists rely on absolute age of rocks and fossils rather than the relative age? Because determining the relative ages of a specimen is often very difficult. Some fossils are hard to identify, especially if the earth movements crush the delicate specimens. Many other fossils and rock layers are eroded away by water and wind leaving no trace of the rock. But for the most part, the rate of radioactive decay in a rock does not change.

There are more than nine methods used to date the radioactivity in rocks—each based on the decay of a certain radioactive element. Usually the first name in the method of dating is the original element. The second name is the end product of radioactive decay. For example, with the potassium-argon method, the isotope potassium decays into argon.

The uranium-thorium dating method is used to determine the age of some of the oldest rocks and meteorites found on earth. Uranium occurs as U-235 and U-238, with U-238 being 140 times more abundant than U-235. U-235 has a half-life of 704 million years. U-

238 has a half-life of 4.5 billion years. Thorium occurs as thorium-232. It has a half-life of 13.9 billion years!

The potassium-argon method of radioactive dating was discovered in 1948 by L. T. Aldrich and A. O. Nier. This method is one of the best because potassium is found in most young and old rocks and is easy to measure. Potassium-40 has a half-life of 1.3 billion years.

The potassium-argon dating method was used to determine the ages of the Hawaiian Islands. The dating showed that the island of Kauai to the far northeast is around 5.6 to 3.8 million years old. Hawaii, with its great volcano Mauna Loa, is less than one million years old!

There are other methods of dating rocks and minerals. The rubidium-strontium method uses rubidium-87, with a half-life of about 47 billion years. Samarium and neodymium are both called rare earth elements. They are used to date very old rocks and meteorites. Samarium has a long half-life of 106 billion years.

Several types of dating methods are used for specific materials. Thorium-proactinium dating is used for ocean sediments. Beryllium dating is often used for the short-term study of ice. Tritium dating, with a half-life of 12.5 years, is used to determine how long it takes for rainwater to travel through cracks in rocks—and has even been used to check the age of wine!

There is one important method for dating organic remains such as wood, parchment, bones, teeth, and skeletons: the carbon-14 method. This type of dating was developed by W. F. Libby in 1947.

Carbon is found in all living organisms and occurs in three isotopes: carbon-12, carbon-13, and carbon-14. Carbon-14 is the most important element when dating organic remains. The half-life of carbon-14 is very short—approximately 5,730 years.

Carbon-14 is constantly being created in the earth's atmosphere. As the sun's rays hit nitrogen gas high in the atmosphere, radioactive carbon-14 is formed. It then combines with the gas oxygen to form carbon dioxide (CO_2)—a common molecule that enters all organisms.

As long as a plant or animal is alive, it contains a balanced proportion of carbon-14. When the organism dies, radioactive carbon is no longer added to the body. As the organism decays, the carbon-14 left in the body gradually gives off beta particles. Using the carbon-14 dating method, scientists can determine the absolute age of the plant or animal!

Carbon-14 dating is important to many branches of science. Paleobotonists, scientists who study the plant life from the past, use carbon-14 dating to determine the age of more recent vegetal remains. Archeologists use this dating method to establish the age of ancient human skeletons. And anthropologists use carbon-14 dating to detect the age of human artifacts.

There are problems with the carbon-14 method of dating. Large portions of the material must be used in the dating process to ensure

Radioactive dating indicates that the rocks at Yosemite National Park are close to 80 million years old.

accuracy. In addition, the method is time-consuming. Also, the carbon-13 isotope often interferes with the detection of carbon-14. Because of the characteristics of carbon-14, this method of dating is reliable only up to about 50,000 years.

Recently, newer methods of carbon dating have been studied, including the use of a cyclotron—an apparatus that separates the carbon-14 from the interfering carbon-13 isotope. Remains as old as 100,000 years have been interpreted from the results!

How do scientists measure the rate of radioactive decay? Radioactive particles are often detected using an instrument called a mass spectrometer. After a specimen of rock is bombarded (for example with a laser), minute particles of the rock are attracted into the mass spectrometer by powerful magnets. The particles are then analyzed by a special measuring device within the mass spectrometer.

Using such intricate instruments in the lab, scientists have been able to determine ages of rocks and fossils to within a few million

The "Genesis" rock from the moon is close to 4.2 billion years old—one of the oldest rocks ever analyzed by radioactive dating.

years or less—a geologically short period of time! Take a rock called quartz monzonite from the tall Half Dome in Yosemite National Park in California. Radioactive dating revealed that the rock is at least 80 million years old. A tree fragment from glacial debris in Green Bay, Wisconsin, has been dated at 11,640 years old. A granite sample from the Transvaal, South Africa, is close to 3.2 billion years old!

Radioactive dating has also been used to help determine the age of rocks from the moon. There are no known fossils on the moon, but by radioactive dating of moon rocks, scientists have determined the ages of major regions of the moon. One of the oldest rocks returned from the Apollo 15 moon mission was the "Genesis Rock"—dated at more than four billion years old! In the future, scientists may be able to visit the other satellites and planets in the solar system. By determining the age of planetary rocks using a radioactive dating method, a time scale for the entire solar system may be developed.

It is easy to see why rocks, minerals, and organic remains containing radioactive elements are called "geological clocks." Radioactive dating has helped to give scientists an idea of how to arrange the different rocks and fossils into the eras, periods, and epochs on the geologic time scale. Now it is time to see the results of those studies—to step back into the earth's ancient past.

3
Precambrian: The Beginnings

The nebula was gigantic. The bits and pieces of dust and gas were more than two times the size of the present solar system. After millions of years, as the gases traveled in a great circle, gravity took over. Close to six billion years ago, pockets of material began to form—the beginnings of planets and satellites.

Scientists believe that the solar system began in this way. The center of the nebula became the sun. The planets and satellites formed along a plane in varying distances from the sun. The rock-hard planets of Mercury, Venus, Earth, and Mars, called the terrestrial planets, formed the inner solar system. The icy, gaseous planets of Jupiter, Saturn, Uranus, Pluto, and Neptune formed in the outer solar system.

The early sun was very active, shooting solar flares high into space. For a few million years after the formation of the solar system, the solar wind—a continuous stream of particles from the sun's atmosphere—blew through space at around two million miles per hour!

The strong solar wind blew around the dust and gases that surrounded the planets and satellites of the inner solar system. The outer planets were too far away from the sun to have their gaseous atmospheres affected by the solar wind.

At first, all of the planets and satellites were extremely hot. Molten, or liquid, rock covered their surfaces. And as the molten material cooled, rocks and minerals began to form on the planetary surfaces.

Some rocks and minerals formed a thick crust, such as on the earth. Others planets and satellites remained hot and active for a very long time. Some never cooled. Today, Jupiter's satellite Io is constantly sending eruptions of hot sulfuric material high into space.

As the earth cooled, heavier materials such as iron and nickel sank to a central core and lighter materials called silicates formed a hard crust. Today scientists know the core is divided into two layers. The inner core has a diameter of 1,740 miles (2,800 kilometers). The outer core measures 1,305 miles (2,100 kilometers) in diameter. The mantle

The earth from space: one of the major planets that formed in the solar system more than 4.6 billion years ago.

is next. This layer of semi-liquid material is around 1,802 miles (2,900 kilometers) thick surrounding the cores. The crust ranges from 3 to 44 miles (4.8 to 70.8 kilometers) in thickness.

Though the crust hardened, the earth was still dynamic and active. Vast ranges of volcanoes grew, sending ash and dust into the sky and covering the surface with molten rock.

Giant chunks of rock called meteorites fell to the surface, bombarding the entire planet. Some of the meteorites struck the earth with the force of thousands of tons of TNT! Many meteorites were made of ice and gas. These gaseous rocks came from the outer parts of the solar system and may have contributed to water forming on the earth.

Tall clouds of gases spewed from cracks in the earth's crust—

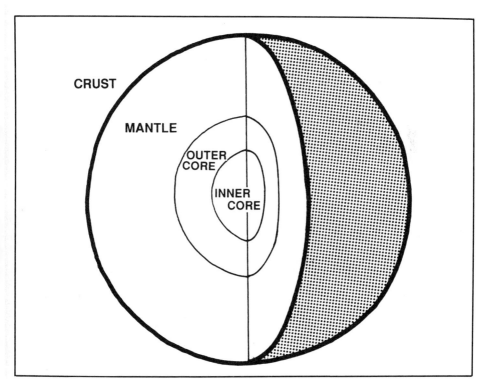

The layers of the earth are broken into the cores, mantle, and crust.

especially carbon dioxide and carbon monoxide. As the earth cooled even more, water vapor began to condense into rain. Additional gases such as methane, ammonia, and nitrogen escaped from the volcanoes. These many gases were the beginnings of the earth's primitive atmosphere.

The activity continued. Great lightning storms charged the atmosphere with electricity, much like the violent lightning storms seen today on the planets of Venus and Jupiter. The gases in the atmosphere reacted with the lightning, forming amino acids, sugars, and other organic compounds. Rains filled giant basins with a salty brine of sulfur, bromine, chlorine, and iodine. This combination of chemicals, amino acids, and sugars, and the warmer ocean basins near volcanic regions eventually helped to encourage the growth of the first organisms on the earth.

Mount St. Helens on May 18, 1980; it may have resembled the volcanic activity on the early earth.

Scientists believe that life grew out of the primitive basins. In 1953, Harold Urey and Stanley Miller, both United States scientists, tried an experiment. First, they filled a miniature chamber with a mixture of gases believed to have been found in the earth's primitive atmosphere. Then they sent electrical sparks through the chamber. These flashes of light, which simulated the lightning storms of the early earth, formed many organic substances in the flask—including amino acids, also known as the building blocks of life!

The time after the formation of the solar system is called the Precambrian era. The Precambrian, or "before Cambrian" (the time before the Cambrian period of the Paleozoic era), represents 87 percent of the earth's history—a little less than 3.9 billion years!

At one time, the entire Precambrian was thought to be a time devoid of all life. But it was known that the rocks were very old. In the mid-1700s, Johann Lehman, a professor of mineralogy (the study of minerals that make up rock) in Berlin, called some of the oldest rocks the "primary series." But few studies were done on the ancient rocks.

PRECAMBRIAN

Million Yrs Ago	Events
-590-	End of Precambrian
-1000-	1st green algae
-2000-	Simple photosynthesis begins
-3000-	Multi-celled organisms grow; considerable volcanic activity
-3500-	1st simple bacteria
-4560-	Earth's crust begins to form

Scientists of the eighteenth century used fossils to determine the age of rock, but they found that some of the older rocks contained no fossils. Therefore, the radioactive dating methods developed in the early twentieth century became important in determining the age of Precambrian rocks.

Scientists have not yet found fossils in the earth rocks that are more than 3.5 billion years old. As a consequence, scientists rely on radioactive dating methods to determine the age of older earth rocks. Some of the oldest have been found in southwestern Greenland. Rocks called Amitsoq gneiss are 3.75 billion years old!

There are only a few places on the earth where ancient rock is found because the earth has an atmosphere and is very active. Wind and water erode the mountains and valleys, and as the active crust of the earth twists and turns, tons of rocks are ground down into sediment. The older rocks have been through a great deal of activity!

In the future, scientists may learn more about ancient rocks by visiting some other members of the solar system. The moon and several other planets and satellites in the solar system are like giant fossils in space. Because they have no atmosphere and their crusts do not move, most of the rocks are just as they were when the planet or satellite was formed!

The fossil record of the Precambrian era is scarce. Some of the fossils are microscopic one-celled animals. Others are no larger than the head of a pin! Most of the early Precambrian fossils resemble single-celled organisms such as primitive bacteria. There were also cyanophytes—organisms that resemble today's blue-green algae.

Many of the bacteria-like Precambrian fossils are located in the mineral chert, a dense form of quartz (silicon dioxide). This rock is often found in rounded nodules or in thin layers. In 1980, scientists found some of the oldest microfossils in western Australia. Under high magnification, scientists found in a piece of chert a simple bacteria—dated 3.5 billion years old!

The primitive cyanophytes were prokaryotic, cells without a

nucleus. (A cell of more recent algae contains a nucleus, needed for reproduction and other life processes such as digestion.) The Precambrian cyanophyte-like fossils were single cells and formed layered structures called stromatolites—rocks that resemble large rounded hassocks! In cross section, ancient and recent stromatolites resemble layer upon layer of wavy lines.

Modern stromatolites are found in isolated regions, especially in shallow water marine environments. The stromatolites are formed when blue-gree algae extract the chemical lime from the ocean water. The algae then deposit the lime in rounded shapes. Scientists believe that the primitive stromatolites formed in the same way, growing in the Precambrian shallow seas. Late in the Precambrian era, the stromatolites grew in abundance—probably because there were no animals to feed on them.

The Precambrian era was a very important time for the first plants. Around two billion years ago, plant cells, the major organisms found on the earth at that time, began a new, startling process: photosynthesis. Plants use photosynthesis to convert carbon dioxide and other elements to carbohydrates—with oxygen as a by-product.

This process led to an abundance of oxygen in the atmosphere and was the start of a major contribution to the earth's living organisms. Some of the oxygen (O_2) in the atmosphere was converted to ozone (O_3)—a layer of oxygen that protected plants from the sun's harmful ultraviolet rays. During the next era, the Paleozoic, this blanket of ozone would play an important part in the growth of land plants.

Toward the end of the Precambrian, there was a new type of life form in the oceans. Single-celled organisms no longer dominated the sea. Multi-celled organisms were beginning to form. One such creature evolved around 1.4 billion years ago—the eukaryotic algae.

The advance of multi-celled organisms was a very important step in the development of the earth. It introduced many new organisms, such as jellyfish and segmented worms. But most of these creatures were composed of soft tissue and no hard parts. Thus, for years, scientists were not able to find fossils of these creatures and plants.

It was a new time for the earth. But the Precambrian era was not only a time for the development of single and multi-celled organisms. It was also a time of change for the earth's continental land masses.

Billions of years ago, the earth's crust broke into sections called plates—much like a giant jigsaw puzzle! Scientists believe that these plates twist and turn because of the movement of the liquid mantle below the earth's crust.

Scientists call this constant rearrangement of the earth's crust plate tectonics. The theory of plate tectonics states that the earth's twelve plates collide, slide under, or move apart from one another—most only inches per year.

One of the most obvious plate collisions is seen at the San Andreas Fault in California. The fault marks the boundary where the North

The bottom layer of the Grand Canyon is one of the oldest rocks in the United States, dating back to the Precambrian era.

American plate is moving south and the Pacific plate is moving north. At unpredictable times, movement of the two plates often creates an earthquake along the fault.

In the Precambrian time, the continents were closer to each other than they are today. Scientists who reconstruct ancient environments believe that the continents were once combined to form a large continent—the first giant landmass on earth. This huge continent eventually broke into sections but not for millions of years.

Today outcrops of Precambrian rock are found on several of the continents. One of the largest is located in northeastern North America, called the Canadian Shield. Here the twisted Precambrian rock stands as a reminder of an ancient era in the Earth's history.

In Africa, Precambrian rocks are found in South Africa, throughout central Africa, in spots along several western countries, and just off the coast in eastern Madagascar. Western Australia is famous for a large deposit of Precambrian rocks, and the tip of India and parts of the coast of Antarctica also contain Precambrian rock.

Precambrian rock contains much of the world's supply of metals and minerals. Gold ranks first as the most important mineral found in Precambrian rock. There are more than a thousand deposits of it—in the deserts of Australia, the mountains of North America, and the famous Witwatersrand rocks of South Africa.

Twenty-five percent of the world's copper comes from a belt of Precambrian rock in Zambia. Precambrian chromite in South Africa provides 50 percent of the world's production. Rich deposits of iron are found in the Canadian Shield, Sweden, and South America. And uranium is mined in the Precambrian rocks of Ontario, Canada, and South Africa.

4

Paleozoic: Signs of Life

The fish was jawless. As it searched for its food, it stayed in shallow water. After all, the eel-like *Sacabambaspis* was a poor swimmer. The rounded head of the creature was protected by bony plates. Its body, covered with thin scales, reached up to 18 inches (45.7 centimeters) in length. What was this strange 470- million-year-old creature? It was one of the earliest known vertebrates or creatures with a backbone. And it was also one of many organisms that lived during Paleozoic times—the Age of Ancient Life.

Most of the creatures that lived during the approximately 370 million years of the Paleozoic era are strangers to recent times. Thick-plated, lobster-like trilobites scurried across the shallow ocean floor in search of food. Scorpion-like eurypterids, some up to nine feet in length, would use their paddle-shaped tails to move along estuaries and lagoons. Some creatures were flat; others were rounded with lumps and bumps.

The first step toward life took place in the latter part of the Precambrian, and by the Paleozoic era, numerous organisms were beginning to flourish—but only in the oceans. There they had plenty of space in which to grow. By the beginning of the Paleozoic, there were only 25 families of shallow water marine animals. But by the middle of the Paleozoic, some 60 million years later, the number had increased to some 150 families of fauna!

The Paleozoic era is divided into seven periods. The early Paleozoic is composed of the Cambrian period, which occurred around 590 to 505 million years ago; the Orodovician period, 505 to 438 million years ago; the Silurian period, 438 to 408 million years ago; and the Devonian period, 408 to 360 million years ago.

The later Paleozoic is broken down into the Carboniferous period, which is further divided into the Mississippian, lasting from 360 to 320 million years ago, and the Pennsylvanian, which spans 320 to 286 million years ago; and the Permian period, the end of the Paleozoic, lasting from 286 to 245 million years ago.

Life dramatically changed on the earth during the Paleozoic era. There was the gradual transition from plants in the oceans to plants on land. The first trees—huge and cone-bearing—developed. New kinds of sea creatures evolved. The number of amphibians, creatures that preferred a combination of water and land, grew. Some of the first reptiles appeared.

The first period, the Cambrian, was a time of tremendous growth. Long before plants started to flourish on land, various types changed and grew in a water environment. There are few records of these plants because they were made of soft tissue, such as seaweed, and left few fossils.

There is some evidence of early plants in the Cambrian sediments—fossil spores. Ancient spores probably behaved much like today's spores produced by fungi and mosses. These plants have no pistils or stamens and must reproduce by sending out spores—tiny particles capable of forming a completely new plant.

The number of marine animals also grew during the Cambrian period. These creatures lived in the warm shallow water of the early oceans and included trilobites, brachiopods, sponges, worms, and other invertebrates, or organisms without backbones. Scientists believe that the increase in marine fauna was due to the increase in shallow ocean areas.

Some animals also flourished because they developed protective shells. Much like the shells of today's turtles and snails, they protected the animals from dangers of the sea, such as attacks from other predatory animals.

One such animal was the trilobite, the name resulting from the creature's three sections—the head, thorax, and tail. There were more than 2,500 different types, ranging in size from microscopic to around 2 feet (a little over 1/2 meter) in length. They were oval in shape, had many pairs of jointed legs, and often had antennae. To protect themselves, certain trilobites would roll into a ball, and fossil trilobites are often found in this position!

Another common Cambrian animal was the brachiopod. These creatures resembled clams, their shells often covered with ridges and

PALEOZOIC

Million Yrs Ago	Periods / Epochs		Events
-245-	Permian		Climate gets colder; many extinctions; ancestors of the 1st dinosaurs appear
-286-	Carboniferous	Pennsylvanian	Coal beds form; 1st reptiles appear
-320-		Mississipian	Amphibians appear; huge land plants; 1st insects
-360-	Devonian		Vertebrates develop; 1st true fishes
-408-	Silurian		1st land plants; photosynthesis
-438-	Ordovician		Age of Invertebrates (animals with no backbone); corals appear
-505-	Cambrian		Growth of marine animals such as trilobites and brachiopods
-590-			

spines. One variety, the *Lingula,* lives today in the ocean around the Philippine Islands—a living fossil from the Cambrian time.

The next period was the Ordovician period, lasting approximately 67 million years. It is often called the Age of Invertebrates, or animals with no backbone. Trilobites and brachiopods were common. New animals also came on the scene. For example, bryozoans, a type of ocean moss, competed with various forms of algae for space on the shallow ocean floor.

Nautiloids, straight or coiled shelled animals with tentacles like an octopus, were some of the largest animals of the Ordovician period. Some nautiloids were close to 12 feet (3.7 meters) in length! Sea lilies, members of the crinoid family, were animals that attached themselves to rocks at the bottom of the sea. They resembled plants and had long, segmented stems. Jawless fish were also abundant, roaming many parts of the oceans in search of food and territory.

The next period, the Silurian, lasted for approximately 30 million

Trilobites were one of the most common animals during the Paleozoic era.

years. This was a major transition period: the first land plants, such as the early fern, began to grow on the shores of ancient seas.

Some scientists believe that land plants did not develop until the Silurian period because of the intense ultraviolet rays of the sun. Unprotected from these dangerous rays, the fragile early plants died.

Since the Precambrian period, the oxygen (O_2) released from plants in the sea was developing into a layer of ozone (O_3). By the Silurian period this layer had become thick enough to block the harmful rays of the sun, and plants began to grow on the shores of lakes and oceans.

Plant development on land was a giant step for plant life. After all, there was plenty of room for plants to spread and survive. In addition, there were few predators to eat the plants.

Another major change took place during the Devonian period, around 360 million years ago: several animals developed an internal skeleton. Animals with skeletons, especially a backbone, are called

The long segmented stems of the crinoids, or sea lilies.

vertebrates. Skeletons developed because organisms were becoming larger and needed the extra support to carry their weight around. As in most large modern animals, such as humans, horses, or whales, skeletons were especially needed to maneuver.

Fish developed during the Devonian period, a time that is also called the Age of Fishes. Most of the Devonian fish did not have true teeth but used sharp ridges to tear apart their prey. Ancestors of the shark also appeared during this time. The modern shark is almost an exact replica of the Devonian shark of almost 400 million years ago!

The late Paleozoic era began with the start of the Carboniferous, or "coal-bearing," period which is broken down into the Mississippian and Pennsylvanian epoch. During the Mississippian, another important step was taken in the earth's long history: amphibians, animals that preferred the land and the sea, developed. It is also called the Age of Amphibians. These animals lived in swampy regions, laying their

Niagara Falls in New York and Canada boasts some of the best examples of Silurian rocks in North America.

eggs in the water. As the eggs hatched, the tadpole-like offspring would swim in the surrounding water. And as the young grew—some up to 6 feet long (1.8 meters)—they eventually lived on land.

Like modern-day rain forests and jungles, the Mississippian swampy forests were ideal for growth—warm water, humid air, and plenty of food for survival. While fishes dominated the sea, the number of amphibians, plants, and insects began to grow on land.

Plants during the Mississipian period were huge. Some were 100-foot (30-meter) plants called scale trees. They and other plants grew to huge proportions along the numerous swamps.

The Pennsylvanian period lasted about 34 million years. Most of the earth's supply of coal was formed during this period. It developed when thick forests of scale trees died in the swamps. They were eventually covered by sediment and after millions of years turned to coal. Other trees also dominated the Pennsylvanian terrain, including many cone-bearing trees and ferns. A type of grass, 40-foot-(12.9-meters)-high rushes, also covered the land.

The Pennsylvanian was also the scene for yet another major event—the appearance of the first reptiles. This had a dramatic effect on the population of amphibians. The reptiles were aggressive animals and would compete with the amphibians for food and territory. Because of the lack of space and food and the growing number of predators, many of the amphibians died out or returned to the water.

There was another reason for the decline of the amphibians: they had to lay their fragile eggs in the water. The reptile eggs were strong—covered with a thick, leathery skin—and the reptiles could lay their eggs on dry land.

During the Permian, the last period of the Paleozoic, many animals disappeared, or became extinct. The climate and the land had changed, with several wet regions becoming dry. In many former swampy areas, amphibians quickly died out. More than 75 percent of all pre-existing species died. But in their place, thousands of other species evolved—especially the reptiles.

Plants of the Permian period were not as affected by the massive extinction. While older types of plants waned, such as the coal swamp floras as the land turned more arid, other types of plants increased and grew in number as they adapted to changing climate conditions.

The growing diversity of plants and animals was not the only activity during the early and late Paleozoic. The continents also became restless, as parts of the earth's surface rose in response to movements of the crust.

Plate tectonics, where large plates of the earth's crust appear to slowly wander around the surface, dominated the Paleozoic era. As the plates traveled across the earth, they collided with other plates. When this happened, mountains rose in response to the collision. In addition, earthquakes and volcanic events occurred.

In the Paleozoic era, great mountain chains arose due to the collision of the plates. Mountains grew in New England, Nova Scotia, and Quebec. The Appalachian Mountains along the east coast of North

Rock layers in the Canyonlands in Utah represent the Pennsylvanian, Permian, Triassic, and Jurassic periods.

America and the Ouachita Mountains of the southeastern United States also rose.

There were times of volcanic activity during the Paleozoic time. Along the east coast of today's North American continent, long rows of volcanoes erupted, sending ash and dust into the atmosphere. In Britain, as the land moved restlessly, mountains were forming and volcanic islands grew. The eruption of Mt. St. Helens in 1980 may have resembled the action of the volcanoes during Ordovician times— high, billowy clouds of dust reaching high into the atmosphere and eventually settling on the surface.

The lowlands and highlands of the continents were also changing. Modern-day Florida was deep under an ocean of salt water. Other areas were covered by smaller freshwater lakes and rivers. Later these areas were lifted up. That is why scientists find outcrops of rock with evidence of marine organisms on dry land.

In the West Guadalupe Mountains, the steep El Capitan promontory stands above the surrounding arid region of west Texas. But during Permian times, the long mountain range was really a coral reef. Today organic reefs form in the warm, shallow seas between 30 degrees North and 30 degrees South latitudes. Scientists believe that the Permian reefs grew under conditions of warm shallow water—a much different environment from that seen in Texas today!

The various types of plants found from Paleozoic times indicate that the continents were also in different positions on the earth. For example, during the Ordovician period, the Sahara Desert was located at the South Pole and covered with a thick layer of ice!

Toward the end of the Paleozoic era, there were numerous changes on the earth. The climate had altered, and many plants and animals had died. The trilobites, some amphibians, and numerous types of corals had become extinct.

The reptiles were able to adapt to the changing climate and started to increase in population. It was the start of something big: the Age of the Dinosaurs.

5

Mesozoic: Middle Ages of Time

Somewhere in the west central part of North America, the giant dinosaur roamed his territory. There was food somewhere, lurking behind a stone, at a river's edge—even under some dense underbrush. The dinosaur, standing more than 18 feet (5.5 meters) in height and weighing close to eight tons, had to eat quite frequently. And some of his favorite choices were found by the river's edge.

The creature was the giant reptile *Tyrannosaurus rex*, one of the last of the carnivorous, or meat-eating, dinosaurs. Because of his great bulk, *Tyrannosaurus* was a formidable enemy. Working his way through the dense undergrowth near the river, he spotted his prey: the leftover kill of a small, horned, duck-billed dinosaur called *Monoclonius*. The small dinosaur was an herbivore, or plant-eater, and enjoyed eating the soft plants that grew along the banks of the river channel. But this time, *Monoclonius* did not watch out for predators.

The large reptile worked his way over a thick cover of plants, scaring the smaller lizards and dinosaurs in his path. A few of the larger creatures turned and challenged the larger dinosaur, but then moved out of his way. Within minutes, *Tyrannosaurus rex* had his meal.

This was only one of many events that occurred during what is called the Mesozoic era ("middle life")—the Age of Reptiles. The Mesozoic era held many changes for the earth. Crocodiles, turtles, and bony fishes became abundant. Strange marine reptiles called ichthyosaurs dominated the seas. And the ancestors of today's insects flourished. Most amphibians became extinct—victims of a changing climate and the spread of the reptiles.

But truly, the most amazing creatures of the Mesozoic era were the dinosaurs. Bones of these massive creatures were not looked at closely until the nineteenth century. Around 1800, Pliny Moody found fossil footprints of dinosaurs in the Connecticut River valley, but he was not able to identify the huge three-toed shapes. William Clark (of Lewis and Clark fame) also found dinosaur bones in 1806 but did not understand their origins.

Mary Anne Mantell, of Sussex, England, was fascinated by fossils she found around her home. In 1822, she discovered a rock that looked as if it had a tooth embedded in it. Her husband sent the bones to Paris to an expert in fossils—Baron George Cuvier. Though Cuvier incor-

This fish, *Xiphactinus*, measuring more than 5 feet (1.5 meters) in length, is from the Cretaceous period.

rectly identified the fossil, he foretold that a new group of animals would arise from the find.

Cuvier was right. By 1842, many more big reptilian bones had turned up in parts of England. Richard Owens, a British scholar, named them *Dinosauria*, from the Greek *deinos* ("terrible") and *sauros* ("lizard").

When dinosaur bones were found in North America, there was a push to find the remains—much like the 1849 Gold Rush to California. In the late 1860s, just after the Civil War, Othniel Charles Marsh and Edward Drinker Cope competed with each other to find the most dinosaur fossils. For almost twenty years, the two scientists would gather enough dinosaur material to keep fossil experts busy for decades!

Dinosaur bones are still being found today. Because of these finds, much is known about the dinosaurs. More than 250 different kinds lived between 225 and 65 million years ago. Their bones are found in

MESOZOIC

Million Yrs Ago	Period / Epoch	Events
-65-		Dinosaurs and several other species of plants and animals become extinct; 1st flowering plants; 1st true birds evolve
	Cretaceous	
-140-		
	Jurassic	Pterosaurs abundant in the air; Ichthyosaurs abundant in the seas
-190-		1st mammals appear
	Triassic	1st dinosaurs appear
-245-		

North and South America, Africa, Australia, Europe, India, China, and Mongolia. Dinosaur footprints—many the size of washtubs—fossil bones, teeth, and dinosaur eggs are some of the remains of the mighty creatures!

Scientists believe that small reptiles called pelycosaurs were the ancestors of the dinosaurs. These lizard-like creatures evolved during the late Permian period of the Paleozoic era. The pelycosaurs were very important: they may have been the first reptiles to "control" their body temperatures—an important step toward survival.

The pelycosaurs were considered cold-blooded animals. Humans are warm-blooded, using internal means to maintain a constant temperature of around 36° (98.67° F). When humans become too hot, the evaporation of sweat cools the body down. When the humans become too cold, internal chemical reactions produce heat.

Cold-blooded animals cannot keep themselves warm or cool by internal means. They must warm up by lying in the sun or cool down by going into the shade. Witness a snake, a cold-blooded animal, warming itself on a rock in the early morning sunlight.

Though pelycosaurs were cold-blooded, they were able to control their body temperatures to a certain degree. The *Dimetrodon*—a large pelycosaur—had a huge fan on his back. When it became too hot, the *Dimetrodon* would raise its fan. The reptile's blood would flow into the fan and cool off the creature!

Thecodonts, reptiles around the size of a collie dog, were probably the next step toward the dinosaurs. Thecodonts, such as the *Rutiodon* of North America and Europe, evolved during the Triassic period. The thecondonts eventually stood up straight, becoming bipedal, or two-legged, animals. These creatures had a great advantage: they developed strong legs that allowed them to run fast—so fast that they could outrun almost every other creature on the earth!

Different groups of thecodonts eventually evolved into the crocodilians, ancestors of today's crocodiles, and pterosaurs, or winged lizards, some of the largest flying animals that ever lived. But the most amazing creatures to evolve from the thecodonts were two groups of dinosaurs—the saurischians ("terrible lizards") and ornithischians ("bird-hipped" because of their resemblance to bones like birds).

The saurischian dinosaurs were either carnivores or herbivores. The most well-known carnivorous saurischians were the 25-foot (7.6-meter) *Megalosaurus,* one of the earliest carnivores; the *Allosaurus,* one of the most powerful dinosaurs, which weighed eight tons, had a massive jaw, and resembled a king-sized kangaroo (but it did not hop); and *Tyrannosaurus rex,* the "king of the tyrant lizards," measuring around 50 feet (15 meters) in length.

The saurischian herbivores were also large. The *Brontosaurus* was about the same size as a blue whale—around 70 feet (21.3 meters) in length and weighing close to 30 tons. The *Diplodocus* grew to 90 feet (27.4 meters) in length. Another large herbaceous saurischian was the *Brachiosaurus*—measuring 40 feet (12.2 meters) tall and weighing at least 50 tons.

The ornithischian dinosaurs were mostly herbivores. They are divided into ornithopods, that ran on long hind legs; ceratopsians, or horned dinosaurs; and ankylosaurs, or armored dinosaurs. The last group of ornithischians were the stegosaurs—many complete with long spikes on their tails. The stegosaur dinosaurs would spear their enemy by whipping around their terrible tail!

The *Shantungosaurus,* an ornithopod from China, reached close to 50 feet (15 meters) in height. The *Stegosaurus,* a stegosaur with a brain the size of an apple, grew to 30 feet (9 meters) and had two long rows of plates on its back. The *Triceratops,* a ceratopsian, measured more than 30 feet (9 meters) in length. The *Ankylosaurus,* a dinosaur with armored plates over its entire body, often grew close to 36 feet (10.9 meters) in length.

Not all dinosaurs were large. The *Fabrosaur,* an ornithischian, grew to only 3 feet (0.9 meter) in length. And the *Stenonychosaurus,* a saurischian, measured close to 2 feet (0.6 meter) in height.

A complete skeleton of a *Brontosaurus* dinosaur, a herbivore that lived during the Mesozoic era.

It is no wonder that the dinosaurs dominated the earth during most of the Mesozoic era. But the dinosaurs were not the only creatures living on the earth at that time. "Winged lizards" called pterosaurs were the first and largest of the flying creatures with backbones. The wingspan of the *Quetzalcoatlus,* the largest known pterosaur, was around 38 feet (11.6 meters)—the length of two buses!

Marine creatures were plentiful. Plesiosaurs, marine reptiles that measured close to 20 feet (6 meters) in length, had a rounded body and a very long neck—like a turtle attached to a snake! The plesiosaurs nested like sea turtles, using their strong flippers to drag their bodies along a sandy beach. After laying their eggs, they would drag themselves back into the water.

One of the most important developments in the Mesozoic era came around 190 million years ago—the first mammals. Most of these early mammals were only as large as a rat. But they would become an important addition to life on the Earth in the next geologic era.

Along with the animals, numerous plant species developed during the Mesozoic era. Large groups of cycads and ferns were common. Ginkgoes, trees with large fan-shaped leaves, evolved. Giant sequoias, much like the redwood trees found in California, found a niche during the late Mesozoic. Magnolias, one of the earliest flowering plants, also became common during this time.

The earth was especially restless during the Mesozoic era. Scientists believed that the earth's internal heating and cooling caused the crustal plates to move across the Earth—the theory called plate tectonics. During the Mesozoic, the crustal plates containing the continents shifted, twisted, and turned in many directions.

During the Triassic period, the continents were pieced together like a giant, finished jigsaw puzzle! This "supercontinent" was called Pangaea. After a short time, Pangaea broke into two pieces: Laurasia and Gondwanaland. Laurasia moved to the north, and Gondwanaland traveled to the south. The two smaller continents eventually broke into pieces, moving to the present positions of the continents.

The movement of the plates also created several new mountain ranges. When plates moved they would often collide. Mountains were formed as two plates collided—pushed up by the great force. In the western United States, the ancestral Sierra Nevada Mountains and Rocky Mountains formed during the late Mesozoic. The size of the seas also changed as the continents shifted. The oceans covered much of the western boundaries of North and South America and much of southern Europe.

Many areas also experienced volcanic explosions and lava flows during the Mesozoic. One famous example is the Palisades along the Hudson River. This thick layer of dark rock formed around 200 million years ago when molten (or liquid) rock erupted from deep below the earth's surface. As the molten rock cooled, it solidified. Other volcanoes were erupting along the coast of western Canada, in Idaho, and along most of Baja California.

Plants and animals were important to each other during the Mesozoic era's 180 million years. Meat-eating dinosaurs and other creatures kept the population of other animals down, and plant-eating animals kept the plants in check.

It was a balanced environmental system. But then, around 65 million years ago, at the end of the Cretaceous period, something went wrong. Many plant and animal species vanished. Within a million years, more than 50 percent of all species that lived on the earth, including the dinosaurs, became extinct.

One theory is that a great meteorite, comet, asteroid, or a number of space objects measuring close to 1 mile (1.6 kilometer) in diameter, hit the earth. Such a violent strike would send a cloud of dust and debris into the air. Plants need sunlight to live, and animals need the sun to keep warm. But the resulting dark cloud would block out the much needed sunlight.

Herbivores would no longer have plants to eat. And when the herbivores died out, the meat-eating creatures would also die. This massive breakdown in the food chain, where each organism depends on the other for food, could be the reason for the animal and plant extinctions at the end of the Cretaceous period.

There may be other reasons for the death of so many plants and animals at the end of the Mesozoic era. A harsh disease could have spread to many species, causing them to die. The movement of the continents could have changed the areas where the plants and animals lived. Or the increase in volcanic activity could have produced dark clouds of dust—blocking the sunlight and causing major climate changes.

The Cretaceous extinctions were not the first. As far back as the Precambrian era, there is evidence of species completely vanishing. The trilobite is one of the most amazing examples. During the Paleozoic era, trilobites were one of the most abundant creatures, but they completely vanished around 245 million years ago.

Thousands of crawling, wriggling, or flying creatures have lived and died out on the Earth. And when they disappeared, it was as if they had never existed.

The extinction of the dinosaurs and other species caused a major change on the earth. The large and small dinosaurs would no longer dominate. The end of the Cretaceous period was the time for the smaller, more versatile animals to spread across the earth. It was the end of the giants and the beginning of a new age—the Age of Mammals.

6

Cenozoic:
The Earth Today

The thunderstorm darkened the sky as it rushed through the valley. The upright, bipedal creature hurried from bush to bush to conceal himself. He could barely see through the dense thicket of the forest. But he had often seen the small deerlike animal rummaging for grass among the tall trees, and he knew its habits. If he was careful and quiet, he would catch the animal and bring it back for his meal.

The animal was close. As the bipedal raised his crude spear, the glasslike arrowhead glistened. But the keen eyesight of the animal was too much for the bipedal. The animal twitched its ears and bounded out of sight. The bipedal would have to go on searching. And next time, he would not fail.

This was ancient man hunting for food. Early humans were basically nomadic, wandering the plains or the mountains in small bands. The choice of food was plentiful—from small mammals and reptiles to various forms of vegetation.

The human species has had a profound effect on the latter part of the Cenozoic era and has been the dominant species for less than a million years. But early in the Cenozoic, other animals flourished.

The numerous extinctions at the end of the Mesozoic era had one major effect: they allowed the smaller mammals and reptiles to expand their territories and increase in number. There was also less competition for food.

During the Cenozoic era, many types of mammals adapted to their habitats. Certain mammals from the forests developed limbs for climbing, such as the tree-climbing monkeys. Where fruits were available, mammals like the elephant developed teeth for grinding plant material. Where meat was plentiful, animals such as the cheetah developed sharp teeth for pulling apart and cutting flesh.

The brains, teeth, and limbs of mammals were also evolving, mainly because climate changes were producing several new habitats. Dry areas grew as mountain ranges developed. Lush, lowland forests became grassy prairies. Many mammals had to adapt to new food sources—changing from browsing to grazing animals.

Cenozoic land mammals were never as large as the giant Mesozoic dinosaurs. But many of the earlier Cenozoic animals were larger than today's mammals. The mammoths and mastodons, elephant-like creatures with large tusks, grew to 18 feet (5.5 meters) in height. Bears were more massive than the grizzly. Ancestors of the beaver grew to lengths of more than 7 feet (2 meters). And a huge ground sloth, as heavy as an elephant, would search for foliage 20 feet (6 meters) above the ground!

Some birds were very large. Fifty million years ago, a bird called *Diatrymasteini* stood 6 1/2 feet (2 meters) high. During the Miocene epoch, the *Osteodontornis*, with a wingspan of 17 feet (5.2 meters), flew on the shores of California. In Argentina during the Pliocene epoch, the *Argentavis*, a bird of prey weighing 265 pounds (120 kilograms) and having a wingspan of 25 feet (7.6 meters), was one of the world's largest birds.

The growth of marine creatures increased during the Cenozoic era. Marine predators such as carnivorous snails and crabs were plentiful. Coral reefs, like those found in today's Carribean Sea, were rapidly growing in warmer waters. Marine mammals such as dolphins increased in number during the middle Cenozoic.

The ancestors of the whales first appeared about 50 million years ago and quickly adapted to the ocean environment. Today's whales are broken into two groups: the toothed whales, originating in the Oligocene epoch, including porpoises and killer and sperm whales; and the whalebone whales, including the plankton-feeding blue, right,

CENOZOIC

Million Yrs Ago		Period / Epoch	Events
- 1-	Quaternary	Recent	Many mammals disappear; glaciers retreat; modern man
- 2.5-		Pleistocene	Ice ages; many large mammals
- 7-	Tertiary	Pliocene	Mammals spread and reach peak in size
- 26-		Miocene	Large marine & land animals
- 38-		Oligocene	1st saber tooth cats
- 54-		Eocene	Several extinctions
- 65-		Paleocene	Great growth of marine and land animals

humpback, and Greenland whales, which first appeared during the Miocene epoch.

Freshwater fishes grew in number. Amphibians such as toads, salamanders, and frogs were relatively abundant. The number of small reptiles, such as turtles, lizards, and snakes, also grew.

By the early Cenozoic, mosses and fungi were abundant, as were cone-bearing plants. The land was soon dominated by flowering plants. By the mid-Cenozoic, the Earth's vegetation was much like today's plant populations with grasslands, prairies, and great forests. Earlier groups of plants were becoming rare, especially ginkgoes and cycads. Today there are no ginkgo trees growing in the wild, and cycads are usually only found in greenhouses.

Cenozoic plant life had a direct effect on the spread and domination of the herbivorous land mammals. Animals traveled from place to place in search of vegetation. And in areas of abundant plant life, many plant-eating mammals grew and flourished.

The giant Harlan's ground sloth was one of the large mammals that lived during the middle of the Cenozoic era.

The first 10 million years of the Tertiary period, during the Paleocene epoch, showed a great growth in marine and land mammals. Ancestral horses, camels, rodents, primitive primates, and bats flourished on land; while whales, seals, and walruses grew in number in the oceans. Extensive deposits of oil were also developed during the Tertiary period. Scientists estimate that 50 percent of the world's oil fields tap Tertiary rocks and more than 38 percent of the rock holds the world's total reserve of oil.

A long period of extinctions occurred during the Tertiary. But thousands of species were not affected. Around 38 million years ago, the climate again began to cool. Several marine organisms, such as a type of plankton called *Gloigerapsis,* and land animals, such as the huge rhinoceros-like animals called *Titanotheres,* became extinct. Scientists do not know the cause of these extinctions.

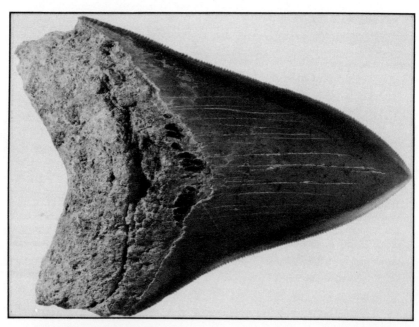

This 3-inch (7.6-centimeter) great white shark's tooth is from the Miocene epoch.

Around two million years ago, the Quaternary period began. It is a time filled with stories of glaciers, animals, and volcanoes. The Quaternary, extending from the present to two million years ago, is often broken into two epochs: the Pleistocene and the Recent (or Holocene). The Pleistocene epoch included some of the earth's coldest periods—the Ice Ages.

During the Ice Ages, the climate changed from warm to cold many times. During the times of extreme cold, many of the continents were covered with great ice sheets. Sometimes the ice was more than a mile thick!

Four great ice advances and retreats occurred. In the United States, the divisions are known as the Nebraskan, Kansan, Illinoian, and Wisconsin. The last retreat during the Wisconsin time occurred between 8,000 and 11,000 years ago.

The Ice Ages had a profound effect on the land and sea during the Pleistocene epoch. Great mountain ranges, like the Adirondack Mountains in New York, were crushed and worn away—the ice sheets grinding mountains down by 650 to 1,000 feet (198 to 305 meters)!

The great ice sheets also changed the courses of rivers and streams. Lakes were carved deeper and new ones formed. It is hard to believe, but there were no Great Lakes in North America before the Pleistocene epoch!

Animals and plants were also affected by the cooling of the climate and the movement of the ice. Many animals, such as reindeer and mammoths, developed thick fur coats to keep warm. Other animals migrated to warmer areas.

Plants "migrated" by sending pollen and seeds to warmer climates—much like milkweed seeds that travel in the late autumn winds. Many hardwood forests were reduced but still survived the advancing ice sheets. And Arctic flowers and grasses followed the margin of the ice sheets as the cold retreated.

There are many areas that show evidence of great ice and water from the Ice Ages. In eastern Washington, there is a patch of land with large ripple marks, huge dry waterfalls, and deep river channels known as the Channeled Scablands. The giant river formations were carved when a great body of water called Lake Missoula broke through a dam of ice during the Ice Ages. Millions of gallons of water poured out over 15,000 square miles (38,850 square kilometers) of land in less than a day!

Some scientists believe that the Earth is still in the grip of the Ice Ages. They believe that the great ice sheets found in Alaska, Greenland, and Asia are the remnants of the last ice retreat. They believe that in the next few thousand years, the ice will again advance to cover many northern regions of the world!

As the ice sheets disappeared from the northern reaches of the continents, so did the larger mammals. More than 50 North American animal groups had become extinct about 11,000 years ago—the beginning of the Recent epoch.

Cycad plants were abundant during the early Cenozoic—but today almost all are seen in greenhouses.

The larger animals such as the imperial mammoth, the American mastodon, and beavers the size of bears became extinct all over the world. Even smaller animals, such as certain groups of horses and the great saber-toothed tiger, no longer existed. Again, scientists can only guess what caused the extinction. Several believe that it was a climate change, while others think that early humans may have hunted and killed off numerous species.

Humans have become the dominant animal during the Recent epoch of geologic time. But other species of animals have also grown in numbers in the last 10,000 years. Today we see moose, bear, birds, whales—successful species whose ancestors adapted to the thousands of environmental changes that have taken place since the beginning of geologic time.

The earth was extremely active during the Cenozoic. Early in the Paleocene, the Rocky Mountains continued to grow. During the Oligocene, volcanic activity blanketed Yellowstone Park and the San Juan Mountains with ash and dust. During the Miocene, volcanic activity continued in the south and central Rocky Mountains. And during Recent times, Mt. St. Helens, Mt. Lassen, and Mt. Ranier—volcanic peaks of the Cascade Range—have all erupted.

The earth's plates have also been very active. The Atlantic and Indian oceans became larger as some of the plates moved away from each other. India, once a small continent, collided with Asia, creating the world's tallest mountains, the Himalayas.

The pushing, pulling, and collision of the plates also produced huge cracks, or faults, in the earth's crust. The earth still experiences these movements—especially along today's most active earthquake zones.

Humans have benefited from the long history of the earth especially in the development of plants, animals, and minerals which have helped humans exist through time. And most importantly, the numerous fossils have helped explain a time when strange creatures and plants covered the ancient lands.

There are many more fossils to be discovered and interpretations to be made. Every day scientists who study ancient life are finding new remains and changing their ideas on how the earth's plants and animals evolved. These new discoveries are important—for a step into the past will help to better under stand the earth's future.

Next time someone asks the time, think about the billions of years of history packed into that question. It is now the Quaternary period, Recent epoch. Perhaps in a hundred thousand years, another event will trigger another time period. Time is continuous, and the earth is constantly changing—making a walk through geologic time anything but dull.

The imperial mammoth was an ancestor of the elephant and often grew to a height of more than 13 feet (4 meters).

Glossary

absolute age—The age of rock or fossils based on radioactive dating.

amber— Hardened resin, usually millions of years old, from pine tree sap; often contains ancient insects.

amino acids—One of the group of organic compounds; also known as the building blocks of life.

amphibian—Only partly adapted to land, this class of organisms laid its eggs in nearby water; young have gills but no legs, while adult amphibians have lungs and four legs.

bipedal—An animal that walks on two legs.

carbon-14—Radioactive carbon, formed when cosmic rays react with nitrogen in the upper atmosphere; also used to date fossils and artifacts of less than approximately 50,000 years old.

carnivore—A meat-eating animal.

cast—What remains when a buried fossil dissolves and is replaced by a mineral.

coal—A rock formed by the decay of woody plants.

cold-blooded animal— An animal that cannot control its temperature by internal means.

continental drift—In plate tectonics, where, over millions of years, the continents appear to move across the earth relative to one another.

core—The central hub of the earth made of iron and nickel.

crust—The upper layer of the earth, composed mainly of light silicate materials.

dinosaurs— The "terrible lizards," a group of animals that dominated the Mesozoic era.

epoch—A short division of the geologic time scale and a subdivision of a period.

era—A long division of the geologic time scale, usually including two or more periods.

extinction—The complete disappearance of a animal or plant species; the extinction of flora or fauna may take a few million years—a short period of time on the geologic time scale.

fauna—The term for animals on the earth.

flora—The term for plants on the earth.

fossil—The buried remains of animals and plants that leave impressions or imprints; fossils can be thousands to billions of years old.

geologic time scale—A chronologic scale of the earth's history based on absolute and relative ages of rock and fossils.

half-life—The time needed for a radioactive element to lose half of its radioactivity by decay, measured from seconds to billions of years.

habitat—A place to live, where an animal finds plenty of food and shelter.

herbivore— A plant-eating animal.

Ice Ages—A time when thick ice sheets covered the northern areas of the world; the last Ice Age ended around 11,000 years ago.

index fossils—Distinct fossils found in a certain layer of rock that are used to identify a specific period of geologic time.

invertebrate—An animal without a backbone.

isotope—Any of two or more species of the same element but differing in their number of protons and neutrons.

mammal—A warm-blooded vertebrate animal; the dominant types of species on the earth today.

mantle—A layer of the earth below the crust and above the core.

meteorite—A space object that lands on the earth, ranging in size from the head of a pin to miles in diameter.

mineral—A combination of elements that usually form specific crystal forms; rocks are composed of minerals.

mold—An imprint of a fossil.

molecule—A combination of two or more atoms.

neutron—A particle found in the nucleus of an atom.

nucleus (of an atom)—The central core of an atom composed of neutrons and protons.

nucleus (of a living cell)—The part of the cell that is responsible for reproduction and processing of food.

outcrop—A large section of rock exposed at the surface.

ozone layer—A blanket of oxygen (O_3) that protects the planet from the sun's harmful ultraviolet rays.

paleoenvironment—The ancient environment.

paleontologist—A scientist that studies ancient plant and animal fossils.

paleontology—The study of ancient plants and animals.

period—A division of the geologic time scale, longer than an epoch.

photosynthesis—A process in a plant that converts carbon dioxide, water, and salts to carbohydrates by the action of sunlight on the plant; the by-product is oxygen.

plate tectonics—The idea that the crust is divided into several large plates that move in different directions across the face of the earth.

prokaryotic—Cells without a nucleus.

proton—A positively charged particle found in the nucleus of an atom.

radioactive dating—A method of using the radioactive decay of elements in a rock to determine the age of the rock or fossils.

radioactive decay—The release of radioactivity by elements over a period of time.

relative age—Determining the age of rock or fossils based on their relative position in the rock strata.

reptile— An egg-laying vertebrate; the eggs are laid on land, with the eggs having a leathery, protective shell.

rock—Material made up of minerals.

sedimentary rock—One of the three major rock types, made of sediment from eroded material.

strata—Layers of rock (and the nickname of William Smith!).

supercontinent—A large hypothetical continent also called Pangaea, that eventually broke apart, forming today's continents.

thorax—The mid-sectionof an animal; often used in reference to trilobites.

trace fossils—Fossilized trails and burrows in rock that represent the movements of ancient animals.

vertebrates—Animals with backbones, such as mammals.

warm-blooded animals—Animals that cool off and heat up their bodies internally.

Further Reading

Asimov, I. *Beginnings: The Story of Origins—of Mankind, Life, the Earth and the Universe*. New York: Walker & Company, 1987.

Charig, A. *A New Look at the Dinosaurs*. New York: Facts on File, 1986.

Eicher, D.L. *Geologic Time*. Englewood Cliffs, N.J.: Prentice-Hall, Inc., 1976.

Faul, H. *Ages of Rocks, Planets, and Stars*. New York: McGraw-Hill Book Co., 1966.

Fisher, David E. *The Origin & Evolution of Our Own Particular Universe*. New York: Macmillan, 1988.

Gallant, Roy A. *Fossils*. New York: Franklin Watts, 1985

Lambert, D. *The Field Guide to Prehistoric Life*. New York: Facts on File Publications, 1985.

Reader, John. The Rise of Life. New York: Alfred A. Knopf, 1986.

Simpson, G.G. *Fossils and the History of Life*. New York: Scientific American Books, 1983.

Stanley S.M. *Extinction*. New York: Scientific American Books, Inc, 1987.

Stidworthy, J. *Life Begins*. Morristown, N.J.:Silver Burdett Company, 1986.

Thompson, Ida. *The Audubon Society Field Guide to North American Fossils*. New York: Alfred A. Knopf, 1982.

Index

A
absolute age, 14-15
Age of Amphibians, 37
   Ancient life, 32
   Dinosaurs, 40, 41. *See also*
      reptiles.
   Fishes, 37
   Invertebrates, 35
   Mammals, 49
amber, 5, 6
amphibians, first, 37
animals, giant, 51
atoms, 15-16

B
beavers, giant, 51, 57
Becqueral, Antoine, 14
biological succession,
      principle of, 12
birds, 43
Boltwood, B.B., 18
brachiopods, *10*, 33, 34-35
Brachiosaurus, 5
Brongniart, A., 12
Brontosaurus, 45, *46*

C
carbon-14, 19-21
Cenozoic, 50-58
Channeled Scablands, 55-56

Clark, W., 6, 42
crinoids, 35, 36
Culvier, Baron George, 12, 42
Curie, M. and P., 14

D
de Monet, J., 6
dinosaurs, 42-47

E
earth, layers, 24-25
extinction, 38, 48-49, 54

F
fishes, 37
fossils, 5-13

G
geologic time, 11
Gondwanaland, 47
Guadalupe Mountains, 40

H
half-life, 16-19
herbivores, 45, 48
Herodontus, 6
Hutton, J., 12

I
ice ages, 55-56
index fossils, 13
invertebrates, 35
isotopes, 16

J
jawless fish, 32

L
Laurasia, 47
Lehman, J., 27
life, early, 28-30

M
mammals, first, 47, 50
mammoths, 51
Mantell, M.A., 42-43
mass spectrometer, 21
mastodons, 51
Mesozoic, 41-49
meteorites, 25, 48
Miller, S., 27
molecule, 15-16

N
Nautiloids, 35
neutrons, 16
nucleus, 16

O
ornithischian, 45
Owens, R., 43
ozone, 29, 36

P
Paleozoic, 32-40
Pangaea, 47
pelycosaurs, 44
Petrified Forest National Park, 8

photosynthesis, 29, 36
plants, first land, 36
plate tectonics, 30-31, 39-40
Precambrian, 23-31
pterosaurs, 44, 47

R
radioactive dating, *20*, 21
radioactivity, 14-22
relative time, 11-12
reptiles, first, 38

S
sedimentary rocks, 6-7
sharks, 37
Smith, W., 12
solar system, early, 23-25
spores, 33
stromatolites, 29

T
thecodonts, 44-45
trace fossils, 9
trees, first, 33
trilobites, 33, 34, *35*
Tyrannosaurus rex, 41, 45

U
Urey, H., 27
uniformitarianism, doctrine of, 12

V
vertebrates, 36-37
volcanos, 40

W
whales, 52-53